YOUR KNOWLEDGE HAS VALUE

Bibliographic information published by the German National Library:

The German National Library lists this publication in the National Bibliography; detailed bibliographic data are available on the Internet at http://dnb.dnb.de .

Imprint:

Copyright © 2012 GRIN Verlag, Open Publishing GmbH
Print and binding: Books on Demand GmbH, Norderstedt Germany
ISBN: 978-3-668-06271-9

This book at GRIN:

http://www.grin.com/en/e-book/308190/a-sensorless-virtual-slave-control-scheme-for-kinematically-disslimilar

Enrique del Sol, R. Scott, R. King

A Sensorless Virtual Slave Control Scheme for Kinematically Disslimilar Master-Slave Teleoperation

GRIN Publishing

GRIN - Your knowledge has value

Since its foundation in 1998, GRIN has specialized in publishing academic texts by students, college teachers and other academics as e-book and printed book. The website www.grin.com is an ideal platform for presenting term papers, final papers, scientific essays, dissertations and specialist books.

A SENSORLESS VIRTUAL SLAVE CONTROL SCHEME FOR KINEMATICALLY DISSIMILAR MASTER-SLAVE TELEOPERATION

E. DEL SOL, R. SCOTT, R. KING
Oxford Technologies Ltd.
7 Nuffield Way, OX14 1RL Abingdon – United Kingdom

K. KERSHAW
Remote handling department, CERN
1211, 23 Geneva – Switzerland

ABSTRACT

The use of telerobotic systems is essential for remote handling (RH) operations in radioactive areas of scientific facilities that generate high doses of radiation. Recent developments in remote handling technology has seen a great deal of effort being directed towards the design of modular remote handling control rooms equipped with a standard master arm which will be used to separately control a range of different slave devices. This application thus requires a kinematically dissimilar master-slave control scheme.

In order to avoid drag and other effects such as friction or other non-linear and un-modelled slave arm effects of the common position-position architecture in non-backdrivable slaves, this research has implemented a force-position control scheme. End-effector force is derived from motor torque values which, to avoid the use of radiation intolerant and costly sensing devices, are inferred from motor current measurement. This has been demonstrated on a 1-DOF test-rig with a permanent magnet synchronous motor teleoperated by a Sensable Phantom Omni® haptic master. This has been shown to allow accurate control while realistically conveying dynamic force information back to the operator.

1. Introduction

The man in the loop has been successfully used in remote handling (RH) at JET for many years [1, 2] and is planned to be used as the main way to carry out maintenance at ITER [3, 4] due to the extreme characteristics that make the use of manual operations prohibitive and the likelihood of unexpected events occurring. The RH techniques have been also deeply researched in particle accelerators like CERN [5, 6] and in nuclear facilities [7].

The common denominator of these facilities is the complex geometry of the working environment, sometimes not designed for RH maintenance, the presence of heavy components with small tolerances, restricted access due the narrow ports and huge dimensions. Those drawbacks may seem smaller when adding the hazardous and hard operating conditions of these facilities such as radiation, poor visibility and in the case of fusion reactors, ultra-high vacuum clean conditions, contaminated dust and some level of magnetic field [3].

The remote maintenance of those premises which have successfully applied these techniques to their standard procedures uses specific equipment to allow force feedback in order to achieve a transparent telemanipulation. These control strategies have been widely studied by telerobotic science which tries to connect humans and robots in order to reproduce the operator actions at a distance [8].

The typical telerobotic system is composed by two environments joined by the communication channel, the local environment where the operator/s is located and the remote environment where the robot or manipulator is operating. The operator environment consists of the human machine interface and the control and command and processing algorithms while the remote environment is composed of the telerobot or slave and the algorithms responsible for managing the incoming and outgoing information. Telerobotic science is considered a subclass of teleoperation where instead of a robot with some degree of autonomy, a machine is used to feel, interact and manipulate with the remote environment [9]. The origins of teleoperation systems go back to the first developments by Ray Goertz in the late forties for the U.S. Atomic Energy Commission where the goal was to protect the workers from the radiation while enabling precise manipulation of material [10]. This radiation protective goal which originated teleoperation technology was shared by the origin of the RH term derived from the remote manipulation of activated waste after nuclear reactions.

There are areas where industrial robots are used in these kinds of hazardous environments, but instead of using teleoperation they are controlled in an automated way e.g. the ISOLDE facility at CERN [11] or the coke ovens repair robots used in Sumitono Metal Industries in Japan [7]. In these circumstances it can be difficult to deal with unexpected events and it would be desirable if a RH master manipulator could be used to control these robots when pre-programmed manoeuvres prove insufficient. However these robots are not backdrivable and can not be driven such as Mascot-like manipulators used at JET and at CERN and inherit from the first Goertz's designs [10]. For that reason these types of robot would be much more flexible if forces and torques were known, allowing a haptic master device to control them and conveying force reflection to the operator in order to achieve the remote manipulation in a safe way. Furthermore, if a general haptic master could be used to teleoperate these industrial robots there would not be any necessity of developing a kinematically similar master for each slave. This can be achieved with a dissimilar master-slave bilateral control allowing the Mascot's master, Dexter[1]® or even the Phantom OMNI® to successfully teleoperate an industrial robot.

This paper is organised as follows: section 2 summarises the remote handling issues in the design of control rooms for teleoperating dissimilar master-slave bilateral systems, section 3 covers the radiation tolerance of robotics systems with special attention to force sensors, in section 4 the sensorless virtual slave architecture is explained and the theory behind the sensorless force feedback is discussed. Section 5 details the experimental equipment used in this research while in section 6 we present the results obtained in 1 degree of freedom. Finally in 7 we present the conclusions and future applications of this technique in multiple degrees of freedom.

2. Remote handling control rooms

Inside of the ITER project Oxford Technologies developed the RH standard work cell to carry out any ITER RH tasks as well as a recommended design for the complete layout of the ITER control room. A standard work cell was realised and built in the control room at Oxford Technologies Ltd headquarters in Abingdon, UK. This work cell was equipped with all the hardware and software components required to create a functional work cell, which is able to simulate any ITER RH task. This design was successfully demonstrated during the spring of 2012 with the virtual accomplishment of three typical RH task located in three different locations at ITER [12].

Two main criteria have been determinant in the final design of the standard work cell, the RH capabilities required by the project and the human factors. These two criteria were taken into account in the designing process of this control room and the initial requirements were transformed into the final design through an iterative process resulting in a very refined design. In the Fig. 1 the first stage of the iterative process is shown where two adjacent cells are displayed, each one mounting a standard haptic master placed on the floor by means of a special support to allow the easy transportation of the master from one cell to another.

---

[1] Dexter haptic master is copyright of Oxford Technologies Ltd., Abingdon, UK.

After this stage the mounting was replaced by a straight beam fixed into the floor in order to avoid discomfort to the operator when he has to move himself around the master.

Due the enormous dimension and variety of remote operated components inside the ITER project, it is expected that several modes of operation will be presented and not only one standard work cell is envisaged to control each RH activity but also other modes are expected including parallel mode and co-operative operation.

With the variety of RH activities that can be carried out, many different slaves are expected to be operated. In order to cope with this diversity and to avoid the creation of dedicated work cells for each RH task to be undertaken, several general manipulators able to control each different slave were selected and two possibilities were finally approved: Dexter manipulator manufactured by Oxford Technologies Ltd and Virtuose 6D40-40 by Engineering Systems Technology. There are two possibilities of operation envisaged for manipulation tasks which are divided in one arm manipulation and two arms manipulation. For the first type of operation both masters can be used whereas for the more complex operations only Dexter will be used [13]. These standard arms are able to control different slaves that will be in general kinematically dissimilar creating therefore the necessity of having a dissimilar master-slave algorithm to cope with that variety.

Fig. 1 Two adjacent work cells, at first design stage, separated by a temporary screen wall

3. Radiation tolerance of telerobotic systems

There are two main effects due the radiation impact into the human being, the deterministic and stochastic effects. While the deterministic effect is characterized by the destruction of a huge number of cells and an easily quantifiable effect into the organ, the stochastic effect is characterized by a modification at a cellular level which increases the probability of several illnesses. In order to describe the radiation into biological level, the equivalent dose is expressed in Sievert (Sv) and it is the product of the absorbed dose expressed in Gray (Gy) with the radiation weighting factor. That factor takes values ranging from 1 for X-rays and γ-rays. The threshold dose of the deterministic effects is higher than 0.5 Gy for an acute irradiation and higher than 0.5 Gy per year for a prolonged irradiation for all tissues except the eyes. The semi-lethal dose for an acute whole body irradiation is estimated in 5 Gy while the CERN annual dose limits for workers is between 20 mSv/year and 6 mSv/year depending on the category. CERN's accelerators in operation can produce an intense radiation fields inside the tunnels, mainly near the collimators areas. The ambient dose rates are in the order of 20 to 200 Sv/h and would lead to death in a few minutes for a person being exposed in the tunnels. For that reason, a cooling time is required before starting the manual maintenance of those areas. That cooling time should be in the order of 4 months for time-consuming interventions [m] in order to obtain a residual dose rate of several mSv/h which is significant but not prohibitive for maintenance works when compared with the limits. Inside of the ITER torus and 11 days after shutdown the level of radiation will rise to 450 Sv/h in the middle of the torus while it will reach 267 Sv/h above the cassette dome. If the time is increased up to 4 months after shutdown those last values will decrease very slowly some tens, making these dose rates much bigger than at CERN facility [15]. In accordance with ITER design description document [16] and [17], the typical RH operations such as the replacement of heavy in-vessel components will be carried out in a radioactive environment of 10 kGy/h gamma dose rate, high temperatures ranging from 50 to 200℃, and total gamma dose going from 1 to 100 MGy [18]. Other example of facility emitting ionizing radiation is the JET experimental reactor in Culham, United Kingdom which is the only operational fusion experiment capable of producing energy up to date. The dose rate level at JET facility 137 days after shutdown is 209 µS/h into the port plasma centre which is several orders of magnitude lower that ITER [19].

The first master-slave manipulators were intrinsically tolerant to the radiation due its mechanical nature. Nowadays the huge amount of electronics included in the modern robots and manipulators make them weak under radiation conditions and components like sensors, drives and electronic circuits have increased the sensitivity to radiation of those robots. A study from AEA Technology and SCK/CEN [20] indicates several limit values for the total dose applied to the main robot components without electronics. Those values vary from several tens of MGy for the robot actuators to 1 MGy for electronics for signal communications specially designed to cope with radiation effects. Tolerant CMOS devices, produced mainly for space applications can operate up to total doses of 1 to 10 kGy [21]. The force sensors based in strain gauges should reach 1 MGy in hardened versions [20] which would lead in longevity of approximately 1 month if those sensors were used in the remote manipulators at worst locations at ITER. This situation, although possible, is not ideal and some techniques without using force sensors should be considered in order to achieve a hard-rad manipulator.

Most of the robotics systems used in radioactive environments have been either especially designed with hard-rad criteria or have been the result of an industrial robot modification in order to fulfil some radiation requirements like the examples presented in [20] and in the AREVA facility [22]. In this paper we present a new approach based on a minimum modification of an industrial robot which enables it to perform manipulation tasks in a radioactive environment.

3.1.1 Force sensor for robots and radiation performance

With the increasing performance in manipulator robots and even in humanoid robots playing a fundamental role in the industry as well as in scientific areas, the use of force sensors fundamentally in the gripper has become necessary [23]. These sensors are used to feel the applied force upon the objects where its load will be measured. The most used force and torque sensors for robots are strain gauges based in piezoelectric effects where a Wheatstone bridge circuit is used to measure the resistance variation with the strain which is then exploited to obtain a signal proportional to the input force.

But not every type of sensor is able to be used in a robotic application. Even the most sophisticated sensors which are able to measure forces and torques with 6 dof with a very small noise in the measures have to fulfil the requirements of robotic applications in terms of size, cost and the special issues of each application.

The findings summarized by Keith E. Holbert et al. in [24] during their performance study of commercial off-the-shelf microelectromechanical (MEMS) systems sensors in a radioactive environment, are one of the best and the first studies concerning the radioactivity issues over pressure transducers based on MEMS technology. The sensors which were used in those experiments were Kulite CT-190, XTE-190, XCE-062 and Endevco 7264C and 7270A founding that most of them were able to support in good conditions radiations below 20 kGy while the Endevco 7264 can operate to more than twice of that value. Older studies like [25] indicate the traditional semiconductor strain-gauge pressure transducers can support 10 kGy of gamma radiation with less than 1 % change in sensitivity and in [26] was found than piezoresistive accelerometers exposed to 3 MG still have a satisfactory dynamic performance only with significant changes in unstrained resistance.

Other completely different type of force sensing is illustrated in [27] where a hydraulic manipulator prepared for ITER uses the difference of pressure between each hydraulic chamber in order to calculate the torque exerted in each joint. This hydraulic manipulator is prepared to support the ITER requirements for its operational area of an estimated dose rate of 300 Gy/h and the accumulated dose of 1 MGy.

In the AREVA recycling plant described in [22] a Staubli RX robot was used equipped with a hard-rad ATI force sensor. These new sensors called DeltaRad and ThetaRad Sensors manufactured by the well-known ATI company are prepared to support more than 10 kGy [28], level which meets the requirements of the AREVA facility in terms of radiation tolerance but would not meet the 1 MGy of ITER necessities.

4. Sensorless force feedback approach

The radioactive environments can be significantly different from each other depending on the dose rate emitted and with them the requirements of the manipulators or robots used within them. In order to cope with the less demanding doses it is likely enough to mount a hard-rad force/torque sensor in the robotic gripper increasing with this the total cost of the robot. When the radiation dose rate becomes very high the solution implemented by [27] consists in a hydraulic manipulator based in water with the consequent risk of water leaks. If that risk is wanted to be totally discarded as well as to maintain a solution with components off-the-shelf when the dose-rate increases such as in ITER-like projects, a sensorless approach can be used. This term indicates the indirect end-effector's force and actuator's torques determination without using force and torque sensors. This sensorless approach would avoid replacing every industrial robot based on electrical actuators for a new and relatively high cost hydraulic solution with the disruptions that this change would cause. This method is also applicable when a redundant system is required. If a traditional electrical slave equipped with force sensor is used, it would be convenient to provide the system with a redundant force estimation that allows the manipulator to continue its tasks in case of failure of any sensor when the device cannot be easily removed from inside of the facility.

With the objective of developing a bilateral control with dissimilar master and slave devices the approach called state convergence is implemented allowing each couple of joints (master and slave joints) to be controlled separately. This research also tries to convey force-feedback to the operator by using a force-position based architecture where the force exerted by the master to the operator is based in the composition of the environment forces as well as other forces arising from the restriction of the master movements due the different kinematics. That external force over the robot end-effector will be determined in an indirect way by the sensorless algorithm based on measurements of the current waveform in each actuator. This method, termed sensorless virtual slave, permits the implementation of teleoperation systems based in force-position architecture where the master and slave have different kinematics producing a cost effective solution as well as a radiation hardened approach.

Based in the previous work of [29] where a virtual robot is presented, this work makes further effort into the use of a sensorless modification. The mentioned virtual slave has identical kinematics than the real slave manipulator and approximate dynamics. This approximate dynamic is due the expected errors in the model of the robot which will create variations with respect to reality, effects that have not been modelled or not considered like some frictions in the joints or gears. The joint values of this virtual robot depend on the position and orientation of the end effector of the master.

The dissimilar master-slave algorithm described in the Fig. 2 copes with the differences between master and slave kinematics and transforms the end effector position and pose to the slave joints values.

The slave model is essentially a dynamic model of the slave with modelled actuators and transmissions which is the centre of the sensorless strategy. Receiving the angle of each robotic joint as an input and the current waveform on each actuator it will output the estimated position of the robot end-effector, the pose and the forces and torques exerted against the environment. Those forces and torques calculated will be transformed into force feedback to the operator depending on its dynamics and kinematics.

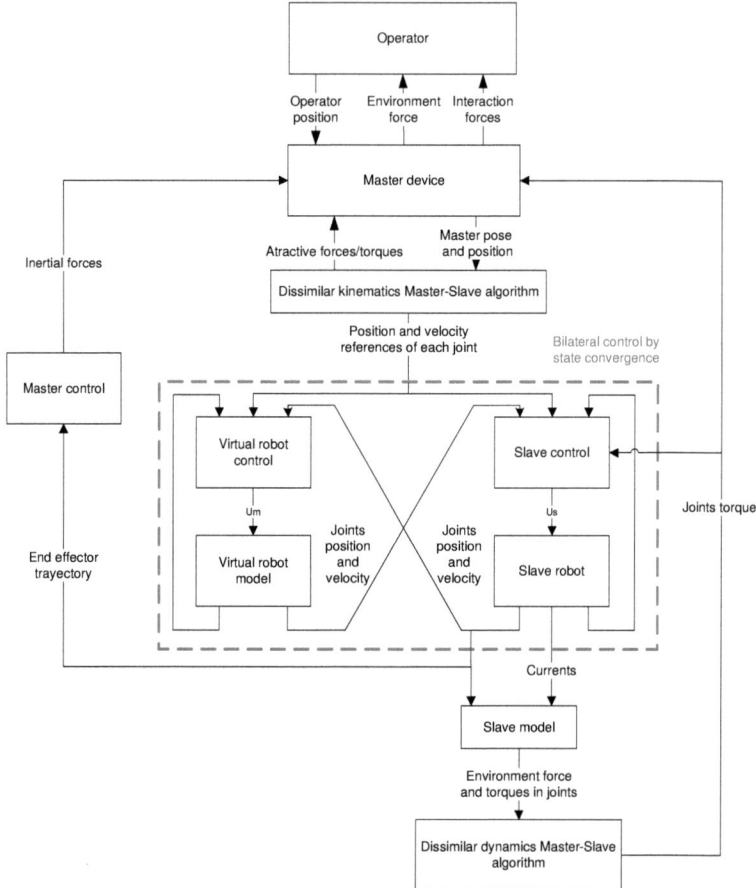

Fig. 2 Bilateral teleoperation system called sensorless virtual slave

4.1 Dissimilar master-slave algorithm

The block diagram represented in Fig. 4 describes in pseudo-code the mentioned algorithm that will be executed in each cycle of the bilateral control. Firstly the position of the haptic master is read and compared with a standard initial position fixed a priori. In order to avoid an abrupt tracking of the master, the initial position and pose of the slave should be correspondent to the master at the starting point. That process will allow the master and slave to converge before starting the bilateral control. Until the haptic-master acquires the mentioned position and pose, the slave will not perform any movement and the forces emitted by the master will be zero. Once the operator reaches the desired point the algorithm will calculate the inverse kinematics of the master continuously in order to detect if the commanded point is inside of the slave's workspace. The inverse kinematics will output the necessary joint values that the slave should have to allow the manipulator's end-effector to reach the position commanded by the master's end-effector. In that case, the manipulator control and the virtual slave control will receive those joints values as a reference. When the result of the inverse kinematics check indicates that the haptic master is pointing in a position or pose which is not reachable by the slave, a force proportional to the distance between the current point and the workspace border will be produced in the haptic device in order to force the operator to return inside the workspace. Also the position of the slave will not be modified. With this method it is possible to control a manipulator with a kinematically different master having a standard initial pose and position which will ensure a safe system.

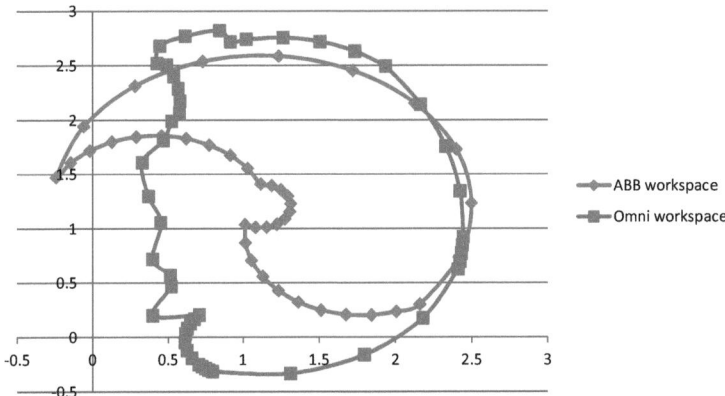

Fig. 3 Overlaid workspaces

In Fig. 3 two overlaid workspaces are displayed. The blue line corresponds to the slave's workspace which is the ABB® IRB 2400/16 industrial robot and the red line corresponds to the Phantom OMNI® workspace scaled and moved to a desired position in order to optimize the common area between them. A kinematic control has been developed with a slave's virtual model to demonstrate the algorithm proposed. In order to implement this algorithm it is necessary to solve the inverse kinematics of the slave controlled by either numerical or analytical methods.

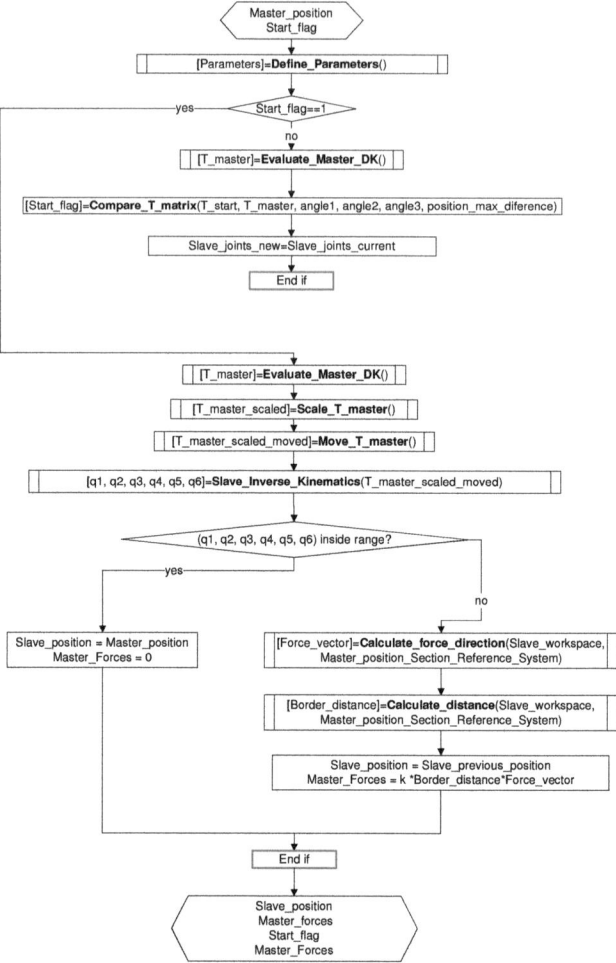

Fig. 4 Dissimilar master-slave pseudo-code

4.2 Master control and inertial forces

The position reached in each bilateral control cycle by the virtual slave will not match in general with the commanded position by the operator due the simulated dynamics properties of the slave. It is possible to represent the dynamic effects of the manipulator producing a drag sensation to the operator. This effect is caused by the inertia and the delays of the real slave in order to acquire the desired position.

The forces felt by the operator can be produced by a composition of three types of effects, the inertial forces, the attractive forces created to push the operator through the slave's workspace and the external forces.
With the proposed system it is possible to control each joint separately which simplifies the control, otherwise an alternative solution like to interpolate homogenous matrix should be used with a previous step of transforming them into quaternions. In order to control each joint a method mentioned in [29] and [30] using space state equations can be possible or on the other hand a classical control method can be implemented.

4.3   Sensorless torque measurement

This research makes use of the current waveform to find out the torque that the actuator is exerting in every moment. There are typically two different methods of measuring the current: via shunt resistor or exploiting one of the several magnetic related effects. These magnetic effects are mainly the Hall Effect principle and the Rogowski principle. The shunt resistor is a very accurate and known resistor which is used to determine the current flowing through the wires via measuring the voltage between its terminals. The Hall Effect is the production of a voltage difference across an electrical conductor transverse to an electric current in the conductor and a magnetic field perpendicular to the current. This effect was discovered by Edwin Hall in 1879 [31]. The operating basis of the Rogowski principle are based in measuring the voltage induced by the change of the current following the well-known expression showing that the induced voltage E is proportional to the rate of change of the current I following (2) [32]. In this method $H = 4\pi \ 10^{-7} \ NA^{-2}$ is the coil sensitivity (V/A). The coil is connected to an integrator, which takes into account the changes on the current value. This integrator can be passive or active or a combination of both and it has a time-constant t which transforms the total gain of the current transducer into the value indicated by (3) [32].

$$V_{out} = R_{sh} \ I \tag{1}$$

$$E = H \ \frac{dI}{dt} \tag{2}$$

$$R_{sh} = H/t \tag{3}$$

The use of current measurements to determine the force exerted by a motor is not new [33]. Some studies use a Hall Effect based current sensor and a low-pass filter to measure the current RMS value. Later an adaptive neuro-fuzzy inference system (ANFIS) was used to correlate the known variables such as current and motor speed with the tool force [34]. The first step in this kind of processes is the training phase then an identification phase can be accomplished. This type of process is used when the function which relates the output variable with the inputs is not known.
Others [33] have used the current measurement to determine a synchronous motor torque with a well-known relationship between torque, current and the angular rotation. The results obtained are relatively good with an error of 6% if a compensation for rotor position is used.
It seems that current measurement is a widely used technique to control actuator torque but more recent modifications of this method have been tested. An example of newer approaches is action/reaction force control which is a technique that exploits one of Newton's principles [36]. This technique is based on taking measures of the disturbance produced in the input of the actuator which could be the current and also other variables like velocity or position in order to determine the torque exerted by the motor.
In this research a Hall Effect sensor was used mainly due the relatively high voltages generated by the drive to power the motor in the order of 300 V which make them impossible to be read with the National Instruments® data acquisition system. While the voltage difference between the shunt terminals would be very low and acceptable in terms of the data acquisition system, the voltage with respect to ground makes it useless for the low voltage data acquisition platform. Other more complex systems based on adapting the

voltage to measurable levels as a differential probe would do were discarded to the high cost for these preliminary tests. The Hall Effect sensors are also less intrusive, cleaner and safer because the current is sensed but not transmitted to the sensor. On the other hand the Rogowski Effect based transducers can present an alternative but their drawback is that they are not able to measure DC components of the current.

4.4 Non-Backdrivable control scheme

The goal of telerobotics is to interface the local and remote environment, creating for the operator the feeling of being in physical contact with the environment. A good telerobotic system should fulfil two basics conditions, i.e., stability and transparency. Stability is essential as a bounded response of the system is an essential requirement for the human operator to be able to perform teleoperation tasks and for safety in particular. While transparency can be achieved when the velocities and forces of the human robot interfaces and the forces applied are the same as the slave ones [8].

A lot of control approaches have been developed to satisfy those criteria depending on the type of the signals transmitted between master and slave. Another classification takes into account the number of channels used for the interconnection between the two sub-systems. This research is focused in the sensorless force feedback of an industrial robot using the current waveform of the actuators to convey forces to the operator. The typical industrial robot used to manipulate heavy loads presents high inertial and frictional properties creating in most cases a non-backdrivable mechanism which will not modify its position in presence of small or medium external forces. This last point results in an elimination of the transparency unless a force-feedback is conveyed to the operator. Most of implementations to produce that feedback use force sensors and transmit the force directly to the operator. However stability problems can arise in systems derived of this architecture and they can be increased even more in presence of time delays [37].

A good approach to overcome those non-backdrivability and stability issues is presented also in [36] where a slave compensator is used receiving as an input the forces sensed by the force sensors and yielding a new slave's position equivalent to the compressed length of a spring supporting the same force. With this approach the slave is safer and the force applied against the external objects is more limited. This method transforms the force measures in a positional error with a local control loop placed in the remote environment. Then a common approach based on passivity control can be used. This procedure also helps to avoid instabilities derived from controlling a slave with dimensions very different from the master in which those differences force the control system to increase the position and force scales [37].

5. Experimental equipment

The master used in these experiments is the Sensable Phantom OMNI® which is capable of providing 6 degrees of freedom of positional sensing with digital encoders (± 5% linearity potentiometers) while is able of exerting torques in 3 degrees of freedom. The continuous exercisable force is 0.88 N and the apparent mass at tip is 45 g. This haptic device is presented in Fig. 5 together with the main equipment included in this research.

An Aerotech® brushless AC motor with its respective drive was used in the experimental setup without any gears, installed in a test bench specially designed for experimental tests. Two current sensors were used to measure the current in to the motor's phases. The most interesting parameters in terms of the election of the current transducer are the motor current ranges which vary from 0 to 8.4 A peak. During some experiments a torque transducer RWT410 series manufactured by Torquesense® (Fig. 6) was mounted into the motor axis capable of measuring torques ranging from 0 to 20 Nm and with a resolution of 0.02% of the full scale which is equivalent to 0.4 Nm with an accuracy of 0.25%. This torque transducer produces an output of ±5 V and can be inserted between the motor and the load measuring the torque exerted by the actuator.

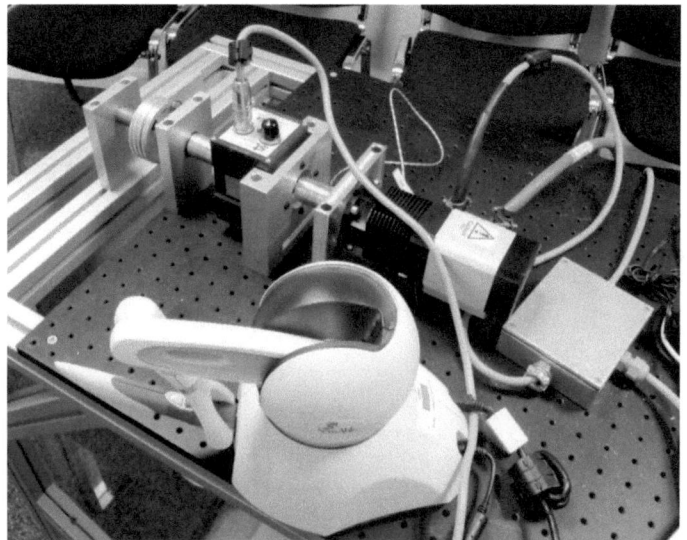

Fig. 5 Experimental equipment

Fig. 6 Torque transducer

The test bench used in this setup is able to support the motor in a free axis movement as well as allow the weight lifting of several weights with a pulley of 32 mm in radius. Two current transducers were used in this research based on the Hall Effect, the ZAP25 manufactured by Amploc® and the TH3A by Multicomp®. The main characteristics of both devices were studied and compared. The data acquisition hardware to capture the information from the current sensor as well as the torque meter used was the National Instruments® NI-USB 6212 which has 16-Bit resolution and it is able to capture 400 kS/s.

Both sensors were previously calibrated and compared to find the exact relationship between the current measured and the output voltage. In order to isolate the sensors from the electromagnetic noisy environment produced by the motor drive the sensors were placed inside a screening box properly grounded. The calibration data was collected commanding several current values to the drive with the motor stall by means of a brake while collecting data from the data acquisition board. The results are showed in Fig. 7 and Fig. 8 with a regression line describing the mentioned relationship. Both graphs show a highly linear function between current and sensor voltage. Phase B was used to monitor the current values and those were acquired using the drive software. It is clear that the sensor TH3A was more convenient for the application intended due its superior sensitivity working with low level currents. The ZAP25 also tend to show a greater content of low frequency noise increasing the inaccuracy of the measurements. For these reasons the TH3A was selected and the following tests were implemented using this sensor.

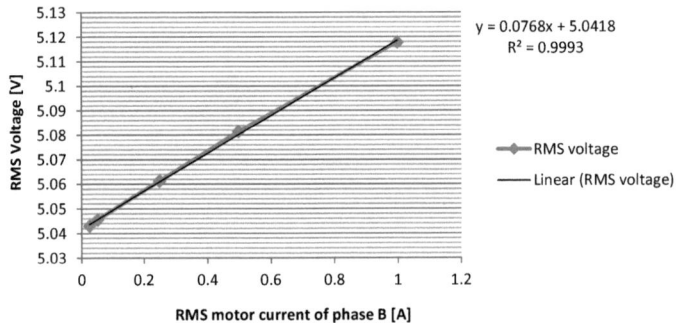

Fig. 7 ZAP25 Calibration results

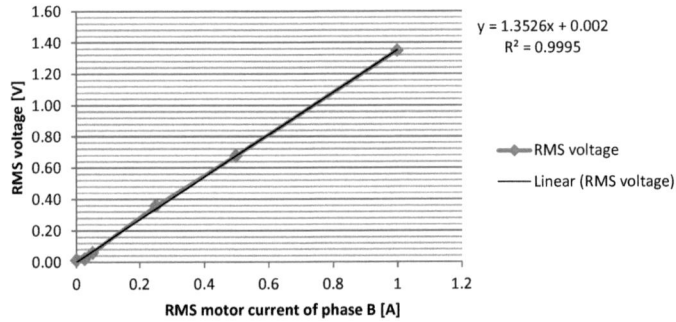

Fig. 8 TH3A calibration results

6. Single degree of freedom results

6.1 Open loop results

With the objective of proving that the current waveform can be used in practice in the determination of the torque exerted by an AC brushless motor like those mounted by most of the modern industrial robots, several trials were carried out in a test bench lifting weights by means of coupling a pulley directly to the actuator without any gears between them. In these tests the current waveform was measured by two different ways, i.e. the Aerotech® drive commanding the motor and the Hall sensor. Both currents waveform results were the same without any significant noise added into the sensor. The torque exerted by the actuator was measured by the torque sensor coupled in the motor's axis in order to check the torque constant provided by the manufacturer. The results of measuring the motor current with the TH3A sensor were extremely good. This was achieved by placing the sensor inside a metallic box in order to avoid electromagnetic noise. In Fig. 9 a graph representing both currents is presented showing one of the worst results obtained when the motor was lifting 1682 g. When the motor is moving the current has a sinusoidal form until the test was terminated and then zero current is measured. Fig. 10 shows the torque measured in the same test, it has significant value until the motor was stopped.

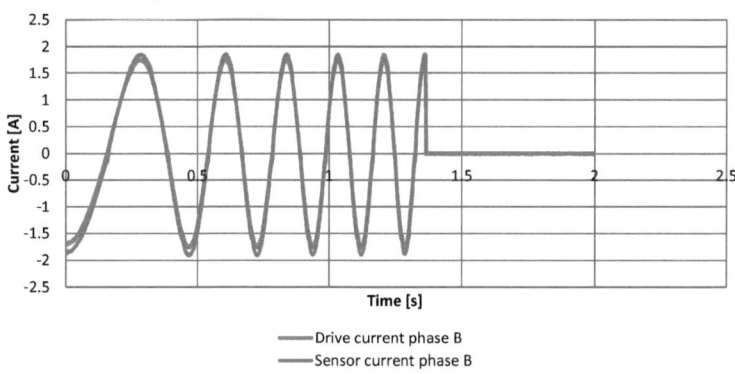

Drive current phase B
Sensor current phase B

Fig. 9. Current waveforms with a load of 1682 g and a current command of 1.75 A

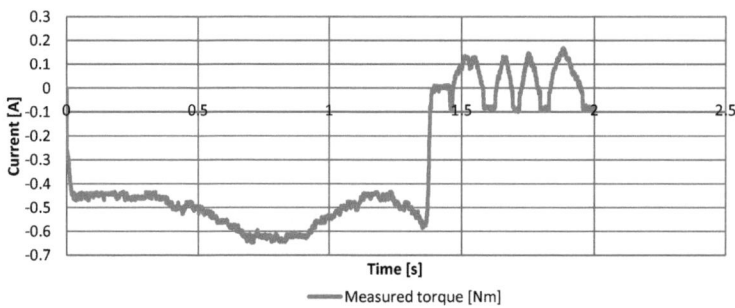

Measured torque [Nm]

Fig. 10. Torque measured with the torque meter

This procedure was repeated with different loads calculating the RMS value of the sensor's current and the torque, after that the torque constant for the RMS value was calculated following the first approximation for the torque given by a motor under a known current which is expressed by the well-known equation (4). The results are presented in the Table 1.

$$\tau_e = K_{e\,RMS} \cdot I_{RMS} \tag{4}$$

Weight [g]	RMS torque value [Nm]	RMS sensor current [A]	RMS Motor constant [Nm/A]
682	0.181145508	0.518627175	0.349278859
1682	0.518533326	1.569251453	0.330433548
2682	0.86292226	1.950905807	0.442318771
3682	1.230992009	2.766710364	0.444929843
4682	1.544131868	3.199304339	0.482646133

Table 1 Experimental motor constant calculation.

These results prove that the instantaneous torque value exerted by a motor can be approximated having the RMS motor constant or the peak constant because they are related with each other assuming sinusoidal waveform of the current if the RMS value of the current is known. The problem is now how to obtain that value in real time allowing a stable teleoperation. In order to achieve a stable and transparent teleoperation the delay should be avoided as far as possible but it still depends on the desired performance. There are two methods of calculating the RMS value of the sinusoidal current applied into a motor, either using one phase and assuming a delay or using two phases without any delay.

Method A.) Using one phase
If a certain delay is admissible and depending on the speed of each actuator it is possible to compute the RMS value of the sinusoidal wave once per period sampling the waveform and applying (7). The period of the current in a brushless motor is related with the electrical speed which is p times the mechanical speed of the rotor, where p is the pair of poles of the motor. In order to calculate the required motor speed it is possible to follow the next equations:

$$\omega_e = \frac{2\pi}{T} \tag{5}$$

$$\omega_m = \frac{\omega_e}{p} = \frac{2\pi}{T * p} \tag{6}$$

Where T is the period of the current waveform, ω_e is the electrical speed and ω_m the mechanical speed. For example, with a maximum acceptable delay of 10 ms, and 4 pair of poles, the resulting minimum rotor speed is 1500 rpm which is a high speed.

$$I_{RMS} = \sqrt{\frac{1}{n}(i_1^2 + i_2^2 + \ldots + i_n^2)} \tag{7}$$

Method B.) Using two phases
With using two phases is possible to obtain the instantaneous peak value without any delay. Let assume the current's waveform described with the following equations:

$$ia = is \cdot \sin(\theta_r + \delta) \tag{8}$$

$$ib = is \cdot \sin(\theta_r + \delta - \frac{2\pi}{3}) \tag{9}$$

Where θ_r is the rotor electrical position and δ the torque angle. These equations can be simplified to:

$$ia = is \cdot \sin(\alpha) \tag{10}$$

$$ib = is \cdot \sin(\alpha - \tfrac{2\pi}{3}) \tag{11}$$

With a new unknown called α which will be determined as follows. Developing the sinus of the sum and replacing is from one equation to the other it is possible to obtain α as:

$$\alpha = atan(\frac{-\sqrt{3}\, ia}{2ib + ia}) \tag{12}$$

And with the value of α replacing it in the (8) we obtain the peak value of the current instantaneously:

$$is = \left| \frac{ia}{\sin(\alpha)} \right| \tag{13}$$

And with the amplitude it is easy to obtain the RMS value of the current:

$$ia\, RMS = \frac{is}{\sqrt{2}} \tag{14}$$

Allowing with it the calculation of the torque in real time with the following expression:

$$\tau_e = K_{e\,RMS} \cdot ia\, RMS = K_{e\,RMS} \left| \frac{ia}{\sqrt{2} \cdot \sin\left(atan\left(\frac{-\sqrt{3}\, ia}{2ib + ia}\right)\right)} \right| \tag{15}$$

6.2 Close loop with force feedback in the Phantom OMNI® haptic master

A bilateral control in one degree of freedom was implemented using the current of two phases to determine the torque produced by the actuator following the method described before. The Phantom® haptic master's joint 2 was used to move the motor axis with both joint angles mapped one to one. The previous means that a 1 rad of movement in the master's joint 2 is equivalent to 1 rad of pulley's turn. The actuator was controlled with a basic PD controller where the outputted control signal was transformed in velocity commands to the motor using LabVIEW® .NET® libraries. The bilateral control loop frequency had values ranging from 10 Hz to 15 Hz.

In the Fig. 11 and Fig. 12 the currents and torque waveform are shown. The currents waveforms were obtained using the Hall Effect TH3A sensors while the torque waveform was obtained using the expression in (15). The starting position of the master corresponds with the weight resting on the floor. It is possible to appreciate in the left part of the Fig. 11 and Fig. 12 how the torque is close to zero and the value of the currents is small before lifting the weight. This effect is better shown in the right part of both figures when the weight has been returned to the floor once it has reached the maximum height. Also between these two areas, the middle part of the graphs shows the movement of the pulley, firstly lifting the weight and secondly dropping it. There is no appreciable difference between these two stages of moving up and down unless the currents' phases are taken into account.

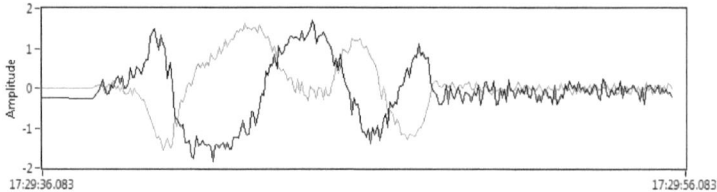

Fig. 11 Currents waveform lifting 682 g weight

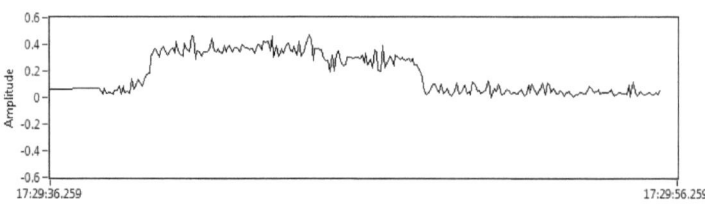

Fig. 12 Torque waveform lifting 682 g weigh

In both objective and subjective ways were possible to detect when the weight got up from the floor and when was deployed by means of force feedback produced into the haptic master.

In doing this test several problems were found that affect to the perceived transparency and stability of the bilateral control. Firstly the frequency of the bilateral loop was to slow to produce a good quality force feedback and even slow to command the actuator in a non-bilateral control. Other important problem found was the correct tuning of the motor controller in order to allow lifting weights without positional error that lead in controller errors. Two states can be distinguished, one is lifting the weight and other would be deploying it. The third problem identified is the current offset. Due the implicit difference in the sensor's behaviour, even small, can introduce an offset in measuring the current and this offset introduced in the equation (15) will lead in a sinusoidal torque with amplitude proportional to the offset.

7.   Conclusions and future work

A one degree of freedom test rig has been developed with the objective of controlling a weight lifting actuator via force-position architecture. The position and velocity difference between the Phantom OMNI® haptic master used in the experiments and the motor´s axis was used to generate a velocity which was used to command the motor´s drive through a LabView® interface. A torque meter was inserted between motor and load with the purpose of verifying the torque´s calculus using the constant of the motor previously calculated. The results have shown firstly that it is possible to use the current of two motor phases to infer the torque produced by the actuator. Secondly, this approach can be used allowing a good enough level of accuracy to render in the operator environment the real forces applied over the motor.   A theoretical framework termed sensorless virtual slave has been proposed to extend this technique towards a 6 dof industrial manipulator. In general, these manipulators are non-backdrivable and that characteristic produces several weaknesses in using these systems for RH. Future developments are focused on improving the bilateral control stability and transparency of the 1 dof control by means of increasing the loop frequency, improving the motor control and implementing a method to overcome the current offset problem. Also there will be new developments in using any non-backdrivable control approach in order to cope with that drawback as well as to extend the sensorless approach into controlling a 6 dof industrial slave.

8. Acknowledgment

This research project has been supported by a Marie Curie Early Stage Initial Training Network Fellowship of the European Community's Seventh Framework Program under the project titled PURESAFE - Preventing hUman intervention for incREased SAfety in inFrastructures Emitting ionizing radiation, project number "264336".

9. References

[1] A.C. Rolfe et al., "A Report on the First remote Handling operations at JET", in: *Fusion Eng. Design*, vol. 46, pp.299–306, Nov. 1999.
[2] O. David and J.P. Friconneau, "Operational experience feedback in JET remote handling" in: *Proc. 23rd Symp. Fusion Tech.*, Venice, Italy, 2004.
[3] A. Tesini and J. Palmer, "The ITER remote maintenance system" in *Fusion Eng. Design*, vol. 83, no. 7, pp. 810–816, Dec. 2008.
[4] T. Honda et al., "Remote handling systems for ITER" in *Fusion Eng. Design*, vol. 63–64, pp. 507–518, Dec. 2002.
[5] R. A. Home et al. "Extended telerobotic activities at CERN," in *ANS Proc. 4th Topical Meeting on Robotics Remote Systems*, pp. 525-534, 1991.
[6] R. A. Horne et al., "Mantis 2 a new long range remote vehicle and servo-master-slave manipulator for the CERN accelerator complex" R.A., CERN, Geneva, CERN-SPS-87-24-(SME), Nov. 1987.
[7] S. Nof, Ed., *Handbook of Industrial Robotics*, Wiley, New York, 1999.
[8] M. Ferre et al., *Advances in telerobotics*, 1 ed, Springer, Berlin, 2007.
[9] T. B. Sheridan, "Teleoperation, telerobotics, and telepresence: A progress report", *Control Eng. Practice, vol.* 3, no. 2, pp. 205–214, Feb. 1995.
[10] P. F. Hokayem and M. W. Spong, "Bilateral teleoperation: An historical survey," *Automatica*, vol. 42, no. 12, pp. 2035–2057, 2006.
[11] E. Kugler, "The ISOLDE facility" in *Hyperfine Interact.*, vol. 129, pp. 23–42, 2000.
[12] B.Haist *et al.*, "TR12 ITER Standard Work Cell – Final Report", Oxford Tecnologies Ltd., Abingdon, Oxfordshire, Report JAC-436-REP-001, Apr. 2012.
[13] D. Hamilton, "RH Control System Standard Parts Catalogue", ITER, Report A6CMLW, Jun. 2012.
[14] M. Brugger, "The radiological situation in the beam-cleaning sections of the CERN Large Hadron Collider (LHC)", Ph.D. dissertation, Inst. Theo. Phys., Graz Tech. University, Graz., 2003.
[15] H. Iida, "Dose Rate Estimate around the Divertor Cassette after Shutdown", ITER Joint Work Site, Max-Planck-Inst. f. Plasmaphysik, Munich, Germany, Analysis Report ITER_D_24VFBN, Oct, 2006.
[16] K. Shibanuma and T. Honda, "ITER R&D: Remote handling systems: Blanket remote handling systems," *Fusion Eng. Design*, vol. 55, pp. 249–257, 2001
[17] EFDA ITER, "Iter design description document - remote handling equipment - wbs23," tech. rep., EFDA, ITER, 2001.
[18] A. F. Fernandez et al., "Toward the development of radiation-tolerant instrumentation data links for thermonuclear fusion experiments," *IEEE Trans. Nucl. Sci.*, vol. 49, no. 6, pp. 2879–2887, Dec. 2002.
[19] R.Villari et al., "Shutdown Dose Rate Benchmark Experiment at JET to Validate the Three-Dimensional Advanced-D1S Method," Fusion Eng. Des., vol. 87, no. 3, 2012.
[20] R. Sharp and M. Decréton, "Radiation tolerance of components and materials in nuclear robot applications," *Reliab. Eng. Syst. Safety*, vol. 53, pp. 291–299, 1996.
[21] A. Holmes-Siedle and L. Adams, Handbook of Radiation Effects, Second Edition, Oxford University Press, Oxford and New York, 2002.
[22] G. Piolain et al. "Dedicated and standard industrial robots used as force-feedback telemaintenance remote devices at the Areva recycling plant*" in Proc. of 1*st* Int. Conf. Applied Robotics for the Power Industry (CARPI)*, Oct., 2010. doi: 10.1109/CARPI.2010.5624418

[23] V. Ganti, "Analysis of 4-dof force/torque sensor for intelligent gripper", B.Tech. thesis, Mech. Eng. Dept., Nat. Inst. of Tech., Rourkela., Orissa, 2011.

[24] K.E Holbert et al., "Performance of Commercial off-the-shelf microelectromechanical systems sensors in a pulsed reactor environment", *Radiation Effects Data Workshop (REDW) IEEE*, Denver, CO, 2010.

[25] F. D. Terry et al., "Transient NuclearRadiation Effects on Transducer Devices and Electrical Cables," Phillips Petroleum Company, Atomic Energy Division, IDO-17103, TID-4500, November 1965, 68 pp.

[26] W. E. Chapin et al., "The Effect of NuclearRadiation on Transducers," Battelle Memorial Institute, REIC report no. 43, TIC report no. 3, 126 pp., October 31, 1966.

[27] P. Nieminen et al., "Water Hydraulic Manipulator for Fail Safe and Fault Tolerant Remote Handling Operations at ITER", *Fusion Eng. Design*, vol. 84, 1420–1424, 2009.

[28] ATI Industrial Automation (2011). *Customized F/T Transducers* [Online]. Available: http://www.ati-ia.com/products/ft/ft_CustomApps.aspx

[29] C. Peña, R. Aracil, and R. Saltaren, "Teleoperation of a Robot Using a Haptic Device with Different kinematics", *International conference EuroHaptics 2008*, Madrid, pp. 181-186, 2008.

[30] J. Azorin et al., "Generalized control method by state convergence for teleoperation systems with time delay," *Automatica*, vol. 40, pp. 1575–1582, Sep. 2004.

[31] E. Hall, "On a new action of the magnet on electric currents," *Amer. J. Math.*, vol. 2, pp. 287–287.

[32] W. F. Ray, C. R. Hewson, "High performance Rogowski current transducers," *Conference Record of the 2000 IEEE Industry Applications Conference*, vol. 5, pp. 3083 – 3090, 2000.

[33] H. Asada et al., "Joint torque measurement of a direct-drive arm," *in Proc. 23rd IEEE Conf. Decision and Control*, Las Vegas, pp. 1332-1337, Dec. 12-14, 1984.

[34] X. Li et al., "Current-sensor-based feed cutting force intelligent estimation and tool wear condition monitoring," *IEEE Trans. Ind. Electron.*, vol. 47, pp. 697–702, June 2000.

[35] I. S. M. Khalil, and A. Şabanoviç, "Sensorless torque/force control", in: *Advances in Motor Torque Control*, 1st ed., Ahmad Mukhtar, 2011, pp. 49-69, doi: 10.5772/862.

[36] P. B Shull and G. Niemeyer, "Experiments in local force feedback for high-inertia high-friction telerobotic systems" *in Proc. of IMECE2007*, © 2007 by ASME, 2007.

[37] P. B. Shull and G. Niemeyer, "Force and position scaling limits for stability in force reflecting teleoperation", *in Proc. of IMECE2008*, © 2007 by ASME, 2008.

YOUR KNOWLEDGE HAS VALUE

- We will publish your bachelor's and
  master's thesis, essays and papers

- Your own eBook and book -
  sold worldwide in all relevant shops

- Earn money with each sale

Upload your text at www.GRIN.com
and publish for free